Local Climate and Energy Program Model Design Guide:
Enhancing Value and Creating Lasting Programs

Climate Showcase Communities
Local Climate and Energy Program

U.S. Environmental Protection Agency | April 2015 | www.epa.gov/climateshowcase

This guide was developed by the U.S. Environmental Protection Agency's Climate Showcase Communities Program,[1] which provided grant funding, technical support, and peer-to-peer information-sharing opportunities to 50 communities around the United States to develop and implement replicable models for local programs to reduce greenhouse gas emissions. This guide draws on the experience and examples of Climate Showcase Communities as they developed innovative models for programs that could be financially viable over the long term and replicated in other communities. Although each local program and its context are unique, we hope the concepts described in this guide—and the examples from Climate Showcase Communities—will be useful to a broad range of local programs, from those just being developed to mature programs looking to refine their current program models.

---

[1] *www.epa.gov/statelocalclimate/local/showcase/index.html*

Contents

This page is intentionally left blank.

Introduction

Around the country, many local governments and their partners are taking up the challenge of implementing programs that help local residents and businesses reduce greenhouse gas emissions and energy use, create jobs, and save money. Although climate change has global impacts, local programs play a key role in reducing the carbon footprint of residents and businesses, and building stronger, healthier communities. Some existing programs were created by an initial investment of public funds (including, but not limited to, economic stimulus funds from the American Recovery and Reinvestment Act) to get new programs off the ground, and test and refine program designs that can be replicated in other communities. As these programs near the end of initial start-up funding—and as other communities embark on new programs—local governments and their partners are evaluating how they will raise revenue and deliver services over the long term—in short, how they will become financially viable.

> This guide was developed for local climate and clean energy program implementers to help create or transition to program designs that are viable over the long term by considering how programs create and deliver value for target audiences and partners, how they raise revenue, and how they can operate cost effectively.

Whether a program is adapting to changing circumstances or just starting out, it is useful to think about a design that will keep the program financially solvent and viable over the long term. This guide was developed for local climate and clean energy (i.e., energy efficiency, renewable energy, and combined heat and power) program implementers to help create or transition to program designs that are viable over the long term by considering how programs create and deliver value for target audiences and partners, how they raise revenue, and how they can operate cost effectively. This guide also has universal applications for basic principles of program design, but is based on experiences with climate and energy programs.

This guide emphasizes strategies for:

- Creating value and turning it into adequate program revenues,
- Developing effective partnerships that leverage each partner's strengths to enhance value and strengthen the bottom line, and
- Delivering a set of services that meet your audiences' needs and align with your organization's strengths and resources.

Specifically, the guide:

- Presents a model framework that can be adapted to fit your program design and guide a structured consideration of alternative options;

- Takes an in-depth look at three focus areas of viable program models:
    - Value creation for program audiences and program revenues,
    - Creating effective partnerships, and
    - Program services;
- Provides worksheets and techniques for analyzing and refining your program model; and
- Describes lessons, case examples, and resources derived from the experience of more than 50 local climate and clean energy programs around the country.

> Local climate and energy programs can increase environmental sustainability and add value to local communities in many ways, including:
>
> - Reducing greenhouse gas emissions,
> - Improving air quality and public health,
> - Creating green jobs and building skills,
> - Increasing home values,
> - Creating stronger communities, and
> - Decreasing energy bills for building owners and tenants.

Although there is no single, prescriptive solution for creating programs that are viable over the long term, this guide provides a structured approach for you to develop an approach for realizing these three strategies by thinking creatively about your program model.

Program Models:
The Foundation of Program Viability

A program model describes how a program creates, delivers, and captures value in pursuit of an overarching goal, such as creating a vibrant local economy with an ever-decreasing carbon and energy footprint. Specific Climate Showcase Communities (CSC) program examples illustrate how they have created value in their communities:

- The Sustainable Transportation for a Sustainable Future program in Salt Lake City, Utah, improves citizens' quality of life by using social media to provide participants with a clear and concise list of choices they can make to reduce their environmental impact and save money. Through the program's Clear the Air Challenge, an online interface tracks the results of participating residents' choices and shares the results with their neighbors. Since the Challenge's inception in 2009, participants have avoided driving more than 5.2 million single-occupant vehicle-miles. To find out more about this program, see its profile (*www.epa.gov/statelocalclimate/local/showcase/sustainable-transportation.html*) and the Clear the Air Challenge website (*cleartheairchallenge.org*).

■ The Whatcom Energy Challenge in Bellingham, Washington, delivers value to residents and business owners through a simple one-stop shop for energy efficiency upgrades, which lower fuel bills and increase the comfort of their homes and workplaces. The program also generates revenue for local contractors performing the work, utilities operating energy efficiency programs, and others providing home energy products and services. By December 2013, the project had tallied 1,013 projects in 768 homes and had 342 participating businesses. To find out more about this program, see its profile (*www.epa.gov/statelocalclimate/local/showcase/whatcom.html*) and the Community Energy Challenge website (*www.communityenergychallenge.org*).

Yard sign used to recognize houses that participate in the Whatcom Energy Challenge in Bellingham, Washington.

■ The Central New York Climate Change Innovation Program works with community partners to build the capacity of New York municipalities to develop climate and clean energy action plans, drawing on the program's knowledge of innovative ideas and political savvy. One partner community recently held a ribbon-cutting event to celebrate the completion of its flagship project—an energy retrofit to the Town's historic 1906 two-room schoolhouse (now used as the Town Hall and Post Office), which completely eliminated the need for a large, aging oil-powered boiler and furnace. To find out more about this program, see its profile (*www.epa.gov/statelocalclimate/local/showcase/central-new-york.html*) and the Central New York Energy Challenge website (*sustainableconnections.org/energy/energychallenge*).

The diagram on the next page illustrates the nine key elements of a program model embodied in these and other programs focused on local climate and clean energy.[2]

---

[2] The program model described in this guide is adapted from a business model framework described in *Business Model Generation* by Alexander Osterwalder and Yves Pigneur (John Wiley & Sons, Inc., 2010).

Program Model

The graphic below illustrates the key elements of a program model. The program's "value proposition" is in the center. The left side illustrates the core back-end infrastructure of the program, and the right side illustrates the public-facing delivery of services. All of these aspects of the program rest on the foundation of costs and revenues illustrated at the bottom of the graphic.

Key Partners	Key Activities	Value Proposition	Audience Relationship	Audience Segments
Who do you work closely with to communicate and deliver services to your audience?	What do you do day-to-day to deliver services?	What are you "delivering" to your audience? What is its value to them?	How do you connect with the values and needs of your audience?	Whose behavior are you seeking to influence?
	Key Resources What are the key assets that help you do your work? (e.g., brand, IT systems, etc.)		**Channels** How are you communicating with your audience about your value proposition? How are you delivering services to them?	

Cost Structure	Revenue Streams
What does it cost to undertake your day-to-day activities and maintain your assets?	Where does the money come from? (And, what can it be used for?)

The right-hand side of the framework describes how your program creates value for your audience by providing a set of services through delivery channels supported by your program's brand or reputation.[3] In turn, your audience provides revenue back to the program either directly or indirectly. Direct revenues may be fees your audience pays to receive program benefits (e.g., homeowners receiving an energy assessment and upgrade or commuters accessing transportation alternatives). Indirect revenues are investments from public and private organizations whose mission and goals are served by the actions your program encourages your audience to take (e.g., emissions reduction, job creation).

The left-hand side of the framework describes the back-end operations of your program, including your partners, the activities you undertake to deliver services, and the resources you and your partners rely on (e.g., computer systems). Together, these elements determine the costs of continuing to conduct your work. The left side of the model provides the infrastructure to support the public-facing aspects of the program on the right side. In short, if the program can continue to deliver enough value so it garners revenues sufficient to cover costs, it will be viable over time.

---

[3] See Appendix E for more information on defining audiences and a list of potential audiences.

The framework can be used to map key elements of programs. The template example on page 6 illustrates how the Sustainable Transportation for a Sustainable Future program in Salt Lake City, Utah, mapped its Clear the Air Challenge program using the program design template. The program focused on a market segment of "fence-sitters" who were amenable to messages about reducing their impact on the environment, but needed incentives to take action. The program's core value proposition was providing this audience (and any others that may be interested) with a focused set of activities to undertake and tools to make it easy to track and gain recognition for undertaking them. It reached its audience through traditional and innovative channels with a message focused on "doing the right thing." The city worked with several stakeholder partners with aligned missions that could provide resources to the program and help it reach its target audience. Given its intensive focus on outreach and behavior change, the program focused its core investments on outreach activities and the development of its brand, website, and social media network. As illustrated in the template, program revenues initially came from federal and non-profit grants, as well as local business sponsorship. Over time, the Salt Lake City Chamber of Commerce—a key partner from the outset—adopted the program and provided it with a new source of funding.

Appendix A includes a template for you to develop a model for your program.

Template Example: Sustainable Transportation for a Sustainable Future (Salt Lake City)

KEY PARTNERS	KEY ACTIVITIES	VALUE PROPOSITION	AUDIENCE RELATIONSHIPS	AUDIENCE SEGMENTS
Who do you work closely with to communicate with and deliver services to your audience?	What do you do day-to-day to deliver services?	What are you delivering to your audience? What is its value to them?	How do you connect with the values and needs of your audience? What are your messages?	Whose behavior are you seeking to influence?
<ul><li>Salt Lake Chamber of Commerce</li><li>Utah Department of Environmental Quality</li><li>University of Utah</li><li>Breathe Utah</li><li>Utah Transit Authority</li><li>Utah Department of Transportation – TravelWise</li><li>Utah Clean Cities</li><li>Air Quality Partner Team</li></ul>	<ul><li>Community outreach</li><li>Social media</li><li>News media</li><li>Program administration</li><li>Partner coordination</li></ul>	<ul><li>Action: Clear, concise action items give residents power in improving air quality.</li><li>Impact: The tracker quantifies personal impact (e.g., gallons of gas, emissions).</li><li>Power: Residents can change their impact and easily share it with their community.</li></ul>	A core message is "doing the right thing": <ul><li>Care for community</li><li>Care for family</li><li>Care for quality of life</li><li>Care for health</li><li>Care for environment</li></ul>	<ul><li>Air "fence-sitters" (i.e., those amenable to messages about reducing their impact on the environment, but needing incentives to take action), including women under 45 with a high school or college degree, who are politically moderate, and who are at home with kids, in school, or in the workforce</li><li>Anyone open to our message (general public)</li></ul>

KEY RESOURCES

What are the key assets that help you do your work? (e.g., brand, IT systems, etc.)

- Clean the Air Challenge (CTAC) brand
- CTAC website
- CTAC social media network

CHANNELS

How are you communicating with your audience about your value proposition? How are you delivering services to them?

- Social media
- Traditional media
- Businesses (employers)
- Email
- Person-to-person

COST STRUCTURE	REVENUE STREAMS
What does it cost to undertake your day-to-day activities and maintain your assets?	Where does your money come from? (And, what can it be used for?)
<ul><li>Website/General Clean the Air Challenge = $18,000 – $30,000 per year</li><li>Staff = $30,000 – $50,000 per year</li></ul>	<ul><li>CSC Grant (done 2012)</li><li>Business sponsorships (Rio Tinto, others)</li><li>Non-profit grants</li></ul>

Evaluating and Evolving Your Program Model

The program model template can help you evaluate your current program model and suggest opportunities for refining it. Questions in red in the diagram on page 8 illustrate the types of questions you can ask to refine your model. For example:

- Where are you creating the most value?
- Can you turn value into revenue?
- Are you providing cost-effective services that deliver the most value to your audience?
- Could others provide some of these services?

Use the program model template in Appendix A to think about these questions for your program and how you might adjust aspects of your program model to add value for your audience in different ways. Later sections of this document will walk you through this process for specific aspects of your program related to revenues, partnerships, and services.

Identifying Lessons Learned and Measuring Program Performance

For existing local climate and energy programs, a key strategy for refining program models is to reflect on what has worked well (and not so well) in the past. Measuring performance and identifying lessons learned from implementation to date will help you evaluate strategies and tactics, and help guide refinements to your program model.

One way programs can evaluate program effectiveness and progress is through performance indicators. Performance indicators measure progress toward program goals and objectives. For more information on performance indicators and how they can be integrated into program design, see Appendix F.

For ongoing programs, you can use the program model to reflect back on program design and operational decisions and lessons learned. For example, you could modify the questions in the diagram to provide a retrospective assessment of your program. Instead of beginning with "What audience(s) should you focus on in the future?," you could ask "What audiences have I focused on in the past who delivered the most value for my program in terms of its goals thus far?" and then use lessons from how you engaged those audiences to define future strategies. The box above provides more information on assessing progress to date using previously established performance measures. Reflecting on successes and lessons learned puts you in a position to look forward. You can also use the program model template to assess how the operating context for your program may be changing and how you should adapt.

Program Model Questions

Key Partners	Key Activities	Value Proposition	Audience Relationship	Audience Segments
Who do you work closely with to communicate and deliver services to your audience?	What do you do day-to-day to deliver services?	What are you "delivering" to your audience? What is its value to them?	How do you connect with the values and needs of your audience?	Whose behavior are you seeking to influence?
• How are partners' needs and roles evolving? • Can they offer additional support for your efforts in the future? • Can they fill any new roles?	• Are you doing the right things? • Could others be doing some of them?	• Where are you creating the most value? • Can you turn value into revenue?	• How can you continue to leverage the relationships you've built?	• What audience(s) should you focus on in the future?

Key Resources

What are the key assets that help you do your work? (e.g., brand, IT systems, etc.)

• What are your most valuable assets?
• How can you leverage that value?

Channels

How are you communicating with your audience about your value proposition? How are you delivering services to them?

• Are there alternate channels to reach your audience?
• Can partners (or others) take over some channels?

Cost Structure

What does it cost to undertake your day-to-day activities and maintain your assets?

• How can you lower costs or increase revenue?

Revenue Streams

Where does the money come from? (And, what can it be used for?)

What are potential funding and revenue streams from:
• Your audience?
• Your program partners?
• Entities in your "market" (e.g., contractors)?
• Other public and private funding sources?
• Others?

The techniques described below can help you think creatively about program design by brainstorming alternative program models.

Prototyping prompts you to ask "What are different ways to deliver services to my target audience(s)?" Starting with your current model, imagine all of the different ways your audience could get the services and value you are currently offering. Consider the following:

- Where would your audience get these services if your program did not exist?

- How could they get the services directly from your partners?

- How would they get these services from other types of organizations?

- If you focused only on the single thing you do best, what would it be? And how would your audience get the other services they need?

Be imaginative (and even unrealistic) in generating an initial list. Once you have exhausted your ideas, go through the list and identify options that seem feasible. Identify what would make these options possible (e.g., different partners, more revenue, a different set of assets).

Storytelling encourages you to think about "What story best describes my program?" and "How does the story change when key elements change?" It is a good strategy for anticipating challenges to your program model, testing how your program would respond, and strengthening its resiliency to an uncertain future.

Starting with your current program model, write down a few sentences describing your current program story from the perspective of a program employee or customer. Now, change a significant element of the story to reflect a possible future event that creates a challenge for the program.

For example, the story for an energy efficiency program customer might go something like this:

My house was always cold in the winter and my heating bills were really high. I saw a yard sign on my neighbor's lawn announcing a program that helps you figure out how to save energy in your home. I called, and they connected me to a great contractor who went through my house top to bottom. She showed me a list of things I could do to save energy and save money. Now, my house is warmer and I can see the savings from my heating bill are already paying for the work.

Now, retell the end of the story if the narrative would have included twists such as this:

... I called, and they connected me to a contractor who missed two appointments, did not explain things well, and left a big mess.

... He showed me a list of things I could do to save energy and save money, but there is no way I could pay that much—even if I ended up saving money over the long term.

How would your program need to change to make sure the story—even with these twists—still resulted in a story you would like to tell? In the case of the unreliable contractor, you may realize you need a customer call center to quickly resolve problems and/or a contractor evaluation and rating system to help your program and customers identify top-performing contractors and discontinue relationships with low performers. In the case of the customer balking at upfront costs, you may realize your program needs to help connect customers with available financial rebates from local utilities—or develop an energy efficiency financing program that provides customers with funds for initial investments that can be paid off over time through energy savings.

Scenarios allow you to answer the question, "How can my program prepare for the unknown?" Starting with your current program model, identify two things about the future that are very important to how your program operates, but are currently uncertain. For example, (1) there is uncertainty about whether demand for services will increase rapidly or slowly, and (2) there is uncertainty regarding the likelihood that grant funding will be available in the future. Place them on a two-by-two axis, such as this one:

Program Scenario Matrix

Level of Demand Services

	Low	High
Low	**Quadrant A:** Low Demand, Low Funding	**Quadrant B:** High Demand, Low Funding
High	**Quadrant C:** Low Demand, High Funding	**Quadrant D:** High Demand, High Funding

Availability of Grant Funding

Next, identify your program's strategy if the future turns out to look like Quadrants A, B, C, or D. Such as the following:

- Quadrant A (low demand, low funding): Shut down or significantly revise goals and approach.

- Quadrant B (high demand, low funding): Pursue fee-for-service strategy.

- Quadrant C (low demand, high funding): Do a near-term push on outreach to drive demand.

- Quadrant D (high demand, high funding): Full speed ahead!

Finally, examine how your program model would need to change to respond to the most likely scenarios. Identify the elements of the model that make you most resilient to future uncertainty and build your strategy around them.

Focus Area: Value Creation and Program Revenues

At a workshop attended by nearly 50 climate and clean energy programs around the country, organizers polled participants about their programs' long-term financial viability. The vast majority recognized that some aspects of their programs would need to change significantly as they transitioned from initial federal grant funding to a program model that was viable over the long term. In an informal poll, most reported they "worried a lot" about their future program design and relatively few said they had "figured it out." What were they most worried about? Nearly everyone in the room said they were worried about where the money will come from.

Without adequate revenues, a program cannot exist for long. *The key to program revenue is value creation.* The more value you can create for your target audiences through your services—and the more effective your delivery channels and partnerships are at building your audience—the more options you have for tapping revenue sources that benefit from the value you create.

> ## Clinton Climate Initiative Home Energy Affordability Loan (HEAL) Program
>
> The Clinton Climate Initiative's HEAL program works with companies to provide their employees with incentives for upgrading the energy efficiency of their homes as an employee benefit. Several communities around the country have adopted the HEAL model as part of their energy efficiency strategies.
>
>
>
> Program profile:
> www.epa.gov/statelocalclimate/local/show case/littlerock.html
>
> Program website:
> www.clintonfoundation.org/clinton-presidential-center/about/heal

The Clinton Climate Initiative's Home Energy Affordability Loan (HEAL) program—which is transforming the market for residential energy efficiency by working with companies to offer home efficiency upgrades as an employee benefit—describes the concept of turning values into revenues in terms of "value pools." These pools are the reservoirs of value that a program builds for different types of audiences and partners. Value pools for the HEAL program are illustrated on the next page.

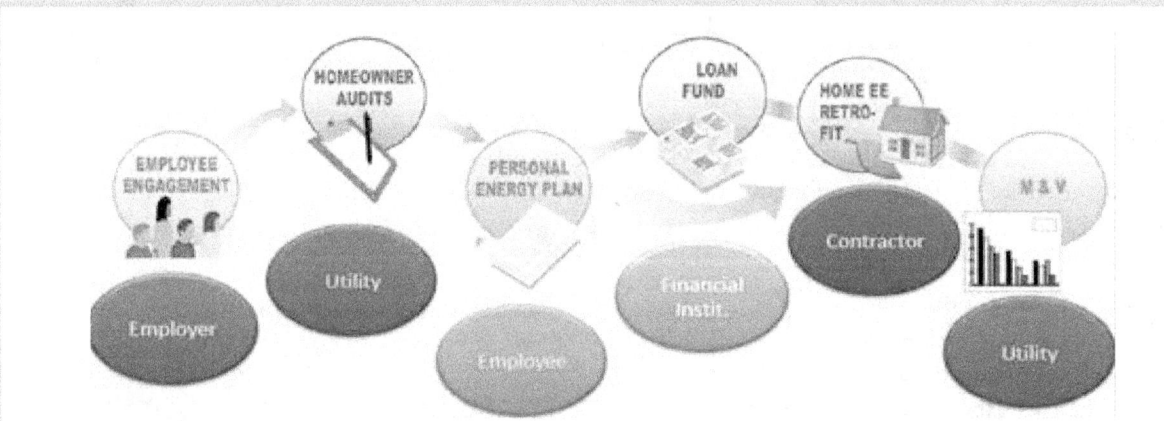

The Clinton Climate Initiative Home Energy Affordability Loan Program uses the concept of value pools to identify how program activities—from employee engagement to measurement and evaluation (M&V)—create value for customers and partners.

Source: Clinton Climate Initiative HEAL Program

As illustrated, the HEAL program creates value pools for:

- Employers, by offering a proven employee benefit model to increase workplace satisfaction and retention;

- Utilities, by highlighting energy savings strategies through home energy audits and helping them meet regulatory targets for energy efficiency with reliable evaluation and data;

- Employees, by helping them identify and implement cost-effective energy efficiency strategies that save money on monthly utility bills and improve home comfort;

- Financial institutions, by creating a fund for employee energy efficiency upgrades that brings in new customers and interest income; and

- Contractors, by marketing energy efficiency assessments and upgrades for them and bringing in new work and motivated customers.

Other types of local climate and energy programs create value pools in many other ways, such as through:

- Emissions reductions,

- Increased capacity (e.g., to develop climate action plans),

- Economic development and job creation,

- Partnerships and relationships,

- Community trust and credibility,

- Improved property value,

- Brand recognition,

- Data and information systems, and
- Replicable program designs.

Turning Value into Revenue

Once you have identified value pools, think about how you can turn value into revenue. Entities that derive value from your program may be willing to pay fees for those services, such as the following:

> ✓ TIP: *Communicate How Your Program Is Creating Value*
>
> When communicating about your program, highlight its economic value—such as how much private investment it is leveraging or how improved air quality increases the economic potential of a downtown area.

- **Customer fees** based on the benefits customers receive from the program, such as energy savings, comfort improvements, or a less stressful commute. In order to overcome customer reluctance to pay upfront fees, the HEAL program has considered using an approach in which customers are not charged any fees until they realize value (e.g., lower energy bills).

- **Contractor fees** based on the benefits contractors receive from your program, such as marketing, work referrals, or quality assurance and customer satisfaction. The HEAL program's experience, for example, suggests energy efficiency contractors value leads at $100–$400 per household. The U.S. Department of Energy (DOE) has found a similar range of residential energy efficiency contractor fees.[4]

- **Financing-based fees** for programs offering loans or other financing for climate and clean energy projects. Like contractor fees, these are based on program services, such as outreach or referrals, which bring business to financial institutions.

- **Employer fees** for programs providing services to employees, such as alternative commuting options, as part of a package of employee benefits.

- **Consulting fees** for programs providing expertise or operational support to other programs or organizations, such as running an energy efficiency program for a local utility.

Your program may also be eligible for public funding to support its pursuit of public policy goals. As an example, the Community Energy Challenge in Bellingham, Washington, received $2 million in funding from the Northwest Clean Air Agency as part of an agreement with a local refinery to partially offset greenhouse gas emissions from the refinery's new low-sulfur diesel fuel production.[5] When seeking public funding, think broadly about the benefits your program provides. For

---

[4] U.S. DOE, Better Buildings Neighborhood Program Business Models Guide: *www1.eere.energy.gov/buildings/betterbuildings/neighborhoods/pdfs/bbnp_business_models_guide.pdf*

[5] For more information, see "Community Energy Challenge Gets $2m in New Funding," *Bellingham Business Journal*: *bbjtoday.com/blog/community-energy-challenge-gets-2m-in-new-funding/24358*

example, climate and energy programs may also be able to seek funding from economic development agencies if they also create jobs, or health agencies if they reduce health hazards.

To fully fund programs, you may need to seek out and assemble many revenue sources. For example, the Whatcom County Energy Challenge calculated the cost of providing its services for home energy upgrades and then analyzed how it could generate sufficient revenue for such a project by assembling revenue from customer fees, referral fees from contractors, carbon mitigation, utility payments, and state economic development funding.

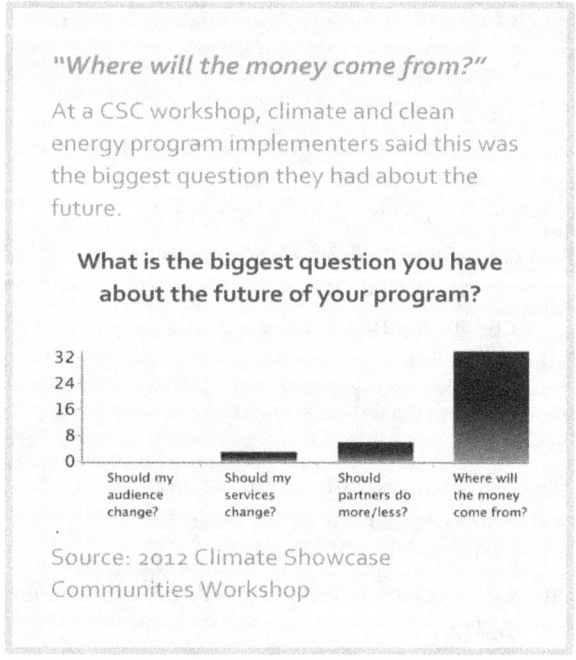

"Where will the money come from?"

At a CSC workshop, climate and clean energy program implementers said this was the biggest question they had about the future.

What is the biggest question you have about the future of your program?

Source: 2012 Climate Showcase Communities Workshop

Value Map

To understand where your program is creating value and how you may be able to tap value for program revenues, fill out a simple value map that asks "Who is benefitting from my program and how?" and "How might they contribute?" A value map will help you identify where you are creating value and suggest ideas for turning value into program revenues. The example below is based on the Clinton Climate Initiative's employer-based energy efficiency HEAL program described above. A blank value map template is included in Appendix B.

Value Map: HEAL Example

WHO IS BENEFITTING FROM MY PROGRAM AND HOW?	HOW MIGHT THEY CONTRIBUTE?
Participating companies: HEAL provides an employee benefit that engages employees and provides marketing and workforce benefits for participating companies.	Participating companies pay a per-employee or flat fee.
Utilities: Residential energy efficiency projects help utilities meet energy efficiency program goals and, in some cases, reduce the need to meet expensive peak demand.	Utilities pay per audit and/or residential upgrade measure.
Financial Institutions: The program generates a demand for loans.	Financial institutions could pay a fee for each loan.
Contractors: The program generates a demand for home assessments and upgrades by providing a stream of motivated and educated customers.	Contractors could pay a fee for program-provided quality assurance services, which increase customer confidence and help drive demand.
Employees: Employees receive quality assured work and financing.	Employer fees indirectly capture the value to the employees.

Resources: Value Creation and Program Revenues

- *U.S. EPA's Clean Energy Financing Programs Website* provides state and local governments with information about different types of clean energy financing strategies, such as rebates, revolving loans, on-bill repayment, energy efficiency mortgages, and others: *epa.gov/statelocalclimate/state/activities/financing.html*

- *U.S. EPA's "Financing Clean Energy Programs" Webinar Series* described how to design and implement funding programs, line up partners, locate available sources of funding, and make clean energy investments more affordable for clean energy program audiences: *epa.gov/statelocalclimate/web-podcasts/local-webcasts.html*

- *U.S. EPA's Financing Program Decision Tool* suggests which types of clean energy financing may be most appropriate for state and local governments based on their target markets and available resources: *epa.gov/statelocalclimate/state/activities/tool.html*

- *U.S. DOE's Better Buildings Neighborhood Program Business Models Guide* provides information about revenue strategies and other aspects of program models for energy efficiency programs: *www1.eere.energy.gov/buildings/betterbuildings/neighborhoods/pdfs/bbnp_business_models_guide.pdf*

- *U.S. DOE's Financing Solutions Center* provides information on a range of financing options and program designs for energy efficiency programs: *www1.eere.energy.gov/wip/solutioncenter/financing.html*

Focus Area: Creating Effective Partnerships

Just as value creation anchors the right, public-facing side of the program model, partnerships anchor the left side as a key element of your program infrastructure. Partners can be an integral part of your program operations by offering program support activities and resources, or can be on the front lines of service delivery and connections with your target audience. Partners can play an important role in revenue generation; they may also be a significant factor in your program costs. It is rare for local climate and energy programs to be successful over the long term without good partnerships. Among other things, these relationships can:

- Extend your program's reach and effectiveness—and can give you credibility with a broader range of audiences;

- Allow the sharing of best practices and fill in gaps in capacity and services; and

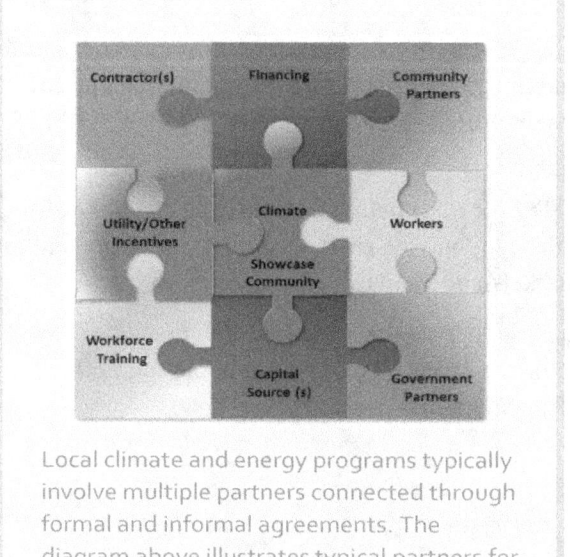

Local climate and energy programs typically involve multiple partners connected through formal and informal agreements. The diagram above illustrates typical partners for programs funded by the U.S. EPA's Climate Showcase Communities Program.

Source: Michael Mann, Cyan Strategies

- Provide access to new audiences, outreach, and educational opportunities—and sometimes new sources of funding.

The puzzle diagram above illustrates the types of partnerships often seen in local climate and energy programs, from contractors and workers rolling up their sleeves for jobs in homes, businesses, and communities, to financial services companies providing the capital to fund the work. As important as the types of partners you have are the ways in which you connect with them and they connect with each other through formal agreements, shared missions, personal relationships, and other connections.

Partnerships are fundamentally about people, and it is important to understand their interests and motivations for coming to the table (e.g., carbon reduction, job creation, business development, community development), as well as their limitations and constraints. Programs have found identifying mutual interests with potential partners and building agreements that address those interests to be successful approaches for building lasting partnerships. For complex and long-term partnerships, it is often important to codify the roles and responsibilities in written agreements. It is important to be aware of common missions that may align organizations in a partnership, but that also create overlaps. Partnerships should focus on each individual partner's strengths and how they can be complementary.

Partnerships and Financial Viability

Partners can play a key role in your program's financial viability by:

- Taking on a larger role in program implementation and funding,

- Providing revenue directly or through their revenue sources, and

- Providing political clout or other leverage for program funding.

For example, the Salt Lake City Clear the Air Challenge implemented its program with more than 25 organizations on its Air Quality Partners Team. As the program matured and the implementing agency sought opportunities to pursue other efforts, it transitioned its Clear the Air Challenge program to one of its key implementation partners—the Salt Lake City Chamber of Commerce. The chamber of commerce now administers the program. Because of its relationships with businesses—including those that operate beyond Salt Lake City—the chamber of commerce has been able to sustain and expand the program.

✓TIP: *Advice for Creating Effective Partnerships*

- Focus on networking and ongoing relationship building (e.g., through in-person meetings and social events).

- Find the right people within an organization to work with.

- Understand partners' goals, skills, and constraints.

- Identify common goals and opportunities.

- Identify what your program offers partners> Help them fill gaps in their capacity and services.

- Know your partners and their concerns.

- Set clear expectations and roles. Formalize commitments and collaboration in writing, if needed.

- Use partnerships to leverage volunteer or pro bono resources, such as universities or utility programs that offer free services from retired engineers.

- For new partnerships, start small and work on one project together to see how effective it is and work out the details on larger collaborative efforts.

As you develop or refine your program model, consider asking the following:

- How are partners' needs and roles evolving? Do changes create opportunities for partners to take on new roles and responsibilities?

- Can existing partners offer additional support for your efforts in the future?

- Can they fill any new roles?

- Should you seek new partners?

- Are you doing the right things? Could partners be doing some of them?

Decisions about whether a program needs new partners (or should retain current ones) can prompt hard questions about program design, such as what services to continue providing, how, and at what cost. Some partnerships may no longer be considered cost effective or necessary to achieve revised program goals.

The City and County of Denver found it needed to ask some tough questions about partners as their revenue strategy shifted from public grant funds to other sources of revenue. Program managers were forced to ask whether they could afford to keep providing certain types of services—and, if these services were to be discontinued, whether they would need to retain the partners who were providing them. Project leads shared the following tips for interacting with partners in similar circumstances:

- Communicate often.

- Be open and candid.

- Share your vision.

- Adapt as needed.

- Understand the political and financial viability of your partnerships (which are not always the same).

✓TIP: *When Establishing Partnerships:*

- Ensure that partners are communicating about your program correctly.

- Recognize that your partners may have varying levels of experience.

- Acknowledge that interest can wane if partners are not seeing the value.

- Anticipate competing agendas or priorities.

- Recognize that well-established programs may not have the flexibility to work with alternative or progressive programs and/or approaches.

- Anticipate that managing partnerships can be time intensive.

Denver Energy Challenge, Colorado

The Denver Energy Challenge focused on upgrading the energy efficiency of commercial buildings in the city. Through its Public Schools Energy Challenge, it also developed an energy efficiency curriculum. The program also helped provide alternative transportation options for city residents.

Program profile: www.epa.gov/statelocalclimate/local/showcase/denver-neighborhood.html

Program website: www.denverenergy.org

A transition to new revenue models and sources can catalyze a new approach to partnerships. Grant-funded programs are often able to attract partners by offering them money. As grant funding declines, programs may need to attract partners who bring their own financial resources, political clout, or strategic leverage. These new partners have their own goals and values, which may influence program design.

It is never too early to think about the future role of partners. Plan for the end from the beginning so you understand how your program can potentially transition program components to other organizations that will carry them forward. In some cases, other organizations may carry on the legacy you began.

Partner Map

To help generate ideas about how partners may play a role in your program (or where their role should change or be complemented by new partners), you can fill out a partner map. One Climate Showcase Communities program that provides energy efficiency upgrade services assessed its partners according to a partner map to help it evaluate whether to maintain current partnerships. The map reflected the following questions:

- Who are existing or potential partners?

- What is their current role in the program?

- What unique opportunities does this partnership represent that are not currently available elsewhere?

- In what areas does the partner excel compared to other partners who could provide these services?

The program's assessment is illustrated in the table below:

PARTNER	CURRENT ROLE	UNIQUE OPPORTUNITIES	AREAS OF EXCELLENCE
Private Company #1	■ Provides advisory services and call center operations for city-wide participants. ■ Provides data evaluation and systems/process updates and improvements. ■ Provides regional services.	Maintains core operations and customer management system. Staffs call center and provides efficiency due to the scale of operations.	Provides services at a lower cost due to the scale of operations. Possesses the flexibility to cover off-and-on tasks.
Non-Profit #1	■ Provides critical outreach in selected neighborhoods. ■ Provides participants for both income-qualifying and non-income-based programs.	Has existing ties to neighborhoods that have traditionally been hard to reach, and possesses bilingual ability.	Ensures city-wide coverage of services.
City Agency #1	■ Provides coordination among various non-profits and channels participants into income-based programs.	Maintains funds for income-based program.	Provides a point of contact for dissemination of program information to city-wide contacts. Manages income-based program.
Non-Profit #2	■ Provides technical advisement. ■ Provides historic property assessments.	N/A	Has the ability to serve as a technical advisor on complicated cases.
City Agency #2	■ Provides grant coordination. ■ Provides program work plans. ■ Provides advisor services. ■ Collectively represents the interests of the group to the public utilities commission and other governing bodies. ■ Provides program evaluation. ■ Provides data management.	Able to receive and manage limited grant opportunities. Has standing in certain governing bodies.	Has the capacity to build consensus with authority.
Non-Profit #3	■ Provides outreach and engagement. ■ Provides marketing services. ■ Provides energy advisement.	N/A	N/A

To develop your own partner map, fill out a similar matrix and then ask the following:

■ What else could this partner do?

■ What other partners might do this work?

A partner map template is included in Appendix C.

Resources: Creating Effective Partnerships

- *U.S. EPA's Local Climate Action Framework, Reach Out and Communicate* page provides information on strategies for engaging the community:
  www.epa.gov/statelocalclimate/local/implementation/communicate.html

- *Climate Showcase Communities Effective Practices Tip Sheets* provide succinct advice from local climate and clean energy program leaders on several types of partnerships:
  www.epa.gov/statelocalclimate/local/showcase/csc-learn.html#tipsheets

  The following tip sheets in particular provide advice on partnerships:

  > Working with Institutional Partners (e.g., other state, local, and regional jurisdictions; public utility commissions; complementary programs)

  > Working with Contractors

  > Working with Students

  > Working with Volunteers

  > Working with Utilities

  > Working with Corporations

  > Identifying and Working with Experts

  > Working Across Ideological Differences

Focus Area: Services

Your program creates value through the services it provides to its audience and partners. For example, the Reading, Riding, and Retrofit project in Buncombe County and Asheville, North Carolina, provides services to its target audience—school administrators, teachers, and students—to help North Carolina schools reduce their environmental footprint. The key services the program provides are:

- Development and maintenance of a concise, online one-stop shop with an easy-to-follow guide that walks school-based green teams through options for making their schools more sustainable and connects them with a large array of resources; and

- Recognition and awards for schools implementing environmental projects based on a point system built into the program website.

Reading, Riding, and Retrofit—Buncombe County and Asheville, North Carolina

The Reading, Riding, and Retrofit project helps schools reduce their environmental footprint through a variety of services, including recognition and project coordination.

Program profile: **www.epa.gov/statelocalclimate/local/show case/reading-riding.html**

Program website: **www.ncgreenschools.org**

Services commonly provided by local climate and clean energy programs include:

- Technical assistance, such as information about how to improve the energy performance of buildings or create a local climate action plan;

- Outreach and communications, such as providing community members with guides for alternative transportation options or other behavior changes to reduce their carbon footprints;

- Recognition, such as yard signs for participating residents or awards for participating organizations; and

- Coordination between customers and service providers, such as connecting homeowners with energy or water efficiency contractors.

The specific services that different audience segments need can vary considerably from community to community. Local climate and clean energy program implementers suggest getting to know your audience well through community meetings or partnering with local organizations before deciding what services to provide. Community partners who work with different audience segments can be an excellent source of information about community needs and can help you identify key services to provide. If you have a diverse audience with diverse needs, your program may want to consider an approach that allows you to customize services for different audience segments. Over time, it is important to listen to your audience and your partners to gauge what services are most valuable and how to deliver them most effectively and efficiently. It is often cost effective to combine service delivery with events or activities organized by others and/or in which your target audience already participates.

Services and Financial Viability

Because services are your primary way of creating value and, in many cases, the primary contributor to your program costs, offering the right services is critical to long-term program viability. Your program can be successful if it provides services that create enough value to generate revenue to cover your costs. It is important to discern what services your audience needs and if yours is the right program to provide them. In some instances, partner organizations may be better positioned to provide certain services. For example, outreach may be critical to your program's goal of changing residents' behavior, but you may want to do outreach through a community partner who has marketing expertise and/or relationships with your target audience. This may then let you focus on what you do best.

Some key questions to ask about services are the following:

- What does your audience need that it is not getting elsewhere? (And, how are its needs changing over time?)
- What services can meet these needs?
- What services can you provide at the highest value and the lowest cost?
- Can partners provide (some of) these services? (And, how can we support them?)

The Cold Climate Community Solutions program in Duluth, Minnesota, is an example of a program that adapted its services to meet emerging needs in its community. The program was

Cold Climate Community Solutions, Duluth, Minnesota

Originally established to encourage homeowners to invest in energy efficiency by helping coordinate homeowners and energy efficiency contractors through a One-Stop Energy Shop, the Cold Climate Community Solutions program added services focused on energy efficient recovery and rebuilding after devastating floods hit the area.

Program profile:
www.epa.gov/statelocalclimate/local/showcase/duluth.html

Program website: *www.duluthenergy.org*

The Duluth Energy Efficiency Program (DEEP) partnered with the city's community action agency to offer a transitional employment program that included a door-to-door canvass so households could see their home's energy losses in infrared and sign up to release utility data. DEEP then provided free home energy score and audit information.

originally established to encourage homeowners to invest in energy efficiency by helping coordinate homeowners and energy efficiency contractors through a One-Stop Energy Shop, as well as through a media campaign and education effort. In June 2012, devastating floods hit the community. Recognizing that many homeowners and businesses would need to rebuild or make repairs, the program focused on ensuring that new buildings were built to use energy efficiently. Leveraging the resources the program had created over time—including relationships with more than 30 local contractors—the program added services focused on energy efficient recovery and rebuilding.

Services Map

Existing programs should periodically review their services to ensure they are meeting community needs and that these services are still the best use of program resources. To help generate ideas about how audiences receive services and who should provide them, fill out a simple services map that asks the following:

- What does your audience need?
- What services can meet these needs?
- Who should provide these services and why?

As needs and/or your program's funding or other circumstances change, it is important to revisit these questions.

The service maps on page 26 illustrate how services—and who provided them—changed as the City of Durham and Durham County, North Carolina, revised its program model from providing direct energy efficiency assessment and upgrade services to homeowners in specific neighborhoods to a broader strategy for raising awareness about energy efficiency and sustainability across the entire city. As shown in the service map for Phase 1 of the program (called the Neighborhood Energy Retrofit Program), the city and county, along with their partners, focused on intensive door-to-door outreach in selected neighborhoods and helped train and coordinate private contractors to do the work. After upgrading several homes but falling short of the program's goals, managers decided to expand the program city wide and focus on broader outreach.

Funds from the U.S. Environmental Protection Agency and DOE for coordinating and incentivizing home energy upgrades were being exhausted after the program successfully upgraded hundreds of homes around the city. Reconstituting as the Home Energy Savings Program, managers shifted to focus primarily on broad outreach to the entire city, including providing outreach in Spanish, relying more on neighborhood associations and "local champions" for outreach, and launching an online social marketing campaign in collaboration with other city departments (as illustrated in the Phase 2 service map on the next page). While contractors continued to provide energy assessment and upgrade services to homeowners, the city and county discontinued its role in coordinating and incentivizing this work. Through this transition, the program's services changed and the program involved new partners to provide them.

Reducing Greenhouse Gas Emissions Through Neighborhood-Based Home Energy Efficiency Upgrades and Capacity Development, City of Durham and Durham County, North Carolina

This program in Durham, North Carolina, was originally designed as a neighborhood-based home energy efficiency upgrade program focused on door-to-door outreach in specific neighborhoods. Over time, its program model and services evolved to focus on broader energy outreach throughout the city and online social marketing.

Program profile: *www.epa.gov/statelocalclimate/local/show case/neighborhood-home-efficiency.html*

The service maps below reflect how the program reconfigured its services and the partners who provide them.

Service Map: Phase 1—Neighborhood Energy Retrofit Program

WHAT DOES YOUR AUDIENCE NEED?	WHAT SERVICES MEET THESE NEEDS?	WHO SHOULD PROVIDE THEM?
Outreach to understand the value and opportunities for energy efficiency upgrades (specific neighborhoods)	Door-to-door outreach in 15 selected neighborhoods	▪ City of Durham and Durham County (program management and coordination) ▪ Trained neighborhood residents to conduct door-to-door outreach, leveraging their role as trusted sources of information ▪ Clean Energy Durham and Advanced Energy, under contract for assistance with program design and implementation
Home energy assessments and upgrades	Assessment and upgrade work	▪ Local contractors provide experience and expertise in home upgrades
	Contractor training	▪ Advanced Energy, under contract to provide on-the-job training to contractors
	Contractor coordination	▪ City of Durham and Durham County
	Quality assurance	▪ Advanced Energy, under contract to provide quality assurance on one-third of upgraded homes

Service Map: Phase 2—Home Energy Savings Program

WHAT DO OUR CUSTOMERS NEED?	WHAT SERVICES MEET THESE NEEDS?	WHO SHOULD PROVIDE THEM?
Outreach to understand the value and opportunities for energy efficiency upgrades (entire city)	Outreach to neighborhoods	▪ Clean Energy Durham, under contract to provide outreach, training, and coordination with neighborhood champions
	Outreach to Spanish-speaking communities	▪ Clean Energy Durham, under contract to provide Spanish-language workshops
	Online social marketing campaign ("Charge Ahead Durham")	▪ City of Durham and Durham County through Home Energy Savings Program, partnering with other city and county departments; provided marketing related to water, waste, and other aspects of sustainability in addition to energy
Home energy assessments and upgrades	Assessment and upgrade work	▪ Local contractors (direct program involvement discontinued)

A services map template is included in Appendix D.

Resources: Services

- *U.S. EPA's Local Government Climate and Energy Strategy Series* describes strategies and services local governments can use to achieve economic, environmental, social, and human health benefits. It covers energy efficiency, transportation, community planning and design, solid waste and materials management, and renewable energy: *www.epa.gov/statelocalclimate/resources/strategy-guides.html*

- *Climate Showcase Communities Effective Practices Tip Sheets* (*www.epa.gov/statelocalclimate/local/showcase/csc-learn.html#tipsheets*) provide succinct advice from local climate and clean energy program leaders on several types of program strategies and services, including the following:

  - Action Checklists
  - Green Teams
  - Incentive Techniques
  - Award/Certificate Programs

- *U.S. DOE's Residential Energy Program Design Guide* provides comprehensive information about residential energy efficiency program services and design: *www1.eere.energy.gov/wip/solutioncenter/pdfs/residential_retrofit_program_design_guide.pdf*

Conclusion

In the past several years, local governments and their partners have driven a flurry of innovation in local climate and energy program design. New programs should consider how they can remain viable over the long term and continue to add value that attracts program investments and other revenues. Existing programs also often need to revisit program models as circumstances change. Key strategies are to:

- Create value and turn it into adequate program revenues;
- Establish effective partnerships that leverage partners' strengths (and your own) to enhance value and strengthen the bottom line; and
- Deliver a set of services that meet audience needs and align with your organization's strengths.

Perhaps the best advice is to listen to local climate and energy program implementers. The same people identified at the beginning of this guide as reporting that their biggest question was "Where will the money come from?" offered advice to their peers that reflected, in their own words, the importance of creating effective partnerships, being ready to adjust services, and creating value:

"You can gain new contacts, knowledge, and networks through working on a project with new partners."

"Don't be afraid to tweak the program if you see that it's not working!"

"If an audience is interested in food, don't sell them on water!"

Appendix A: Program Model Template

The program model template helps you understand and describe how your program creates, delivers, and captures value. Questions in italics illustrate the types of questions you can ask to refine your model. Think about these questions for your program and how you might create or adjust aspects of your program to add value for your audience in different ways.

KEY PARTNERS	KEY ACTIVITIES	VALUE PROPOSITION	AUDIENCE RELATIONSHIPS	AUDIENCE SEGMENTS
Who do you work closely with to communicate with and deliver services to your audience? ▪	*What do you do day-to-day to deliver services?* ▪	*What are you delivering to your audience? What is its value to them?* ▪	*How do you connect with the values and needs of your audience? What are your messages?* ▪	*Whose behavior are you seeking to influence?* ▪
	KEY RESOURCES *What are the key assets that help you do your work? (e.g., brand, IT systems, etc.)* ▪		**CHANNELS** *How are you communicating with your audience about your value proposition? How are you delivering services to them?* ▪	

COST STRUCTURE	REVENUE STREAMS
What does it cost to undertake your day-to-day activities and maintain your assets? ▪	*Where does your money come from? (And, what can it be used for?)* ▪

Appendix B: Value Map Worksheet

To understand where your program is creating value and how you can tap value for program revenues, fill out a simple value map that asks "Who is benefitting from my program and how?" and "How might they contribute?" Through a value map, programs can identify where they are creating value and list ideas for tapping value for program revenues.

WHO IS BENEFITTING FROM MY PROGRAM AND HOW?	HOW MIGHT THEY CONTRIBUTE?
■ ■ ■ ■ ■ ■	■ ■ ■ ■ ■ ■

Appendix C: Partner Map Worksheet

To help generate ideas about how existing partners may play a different role in your program (or where you may want to attract new partners), assess partner opportunities by asking the following questions:

- Who are existing or potential partners?

- What is their current role in the program?

- What unique opportunities does this partnership represent that are not currently available elsewhere?

- In what areas does the partner excel compared to other partners who could provide these services?

- What else could these partners do?

- What other partners could do this work?

Who are existing or potential partners?	What is their current role in the program?	What unique opportunities does this partnership represent that are not available elsewhere?	In what areas does the partner excel compared to other partners who could provide these services?	What else could these partners do?	What other partners could do this work?
. . . . . .	. . . . . .	. . . . . .	. . . . . .	. . . . . .	. . . . . .

Appendix D: Services Map Worksheet

To identify or revisit the services your program should provide, fill out a services map worksheet that asks the following:

- What does your audience need?
- What services can meet these needs?
- Who should provide these services and why?

When considering whether you should provide services directly or partner with others to provide them, consider your program's capacity, skills, and costs relative to those of partners who may provide similar services. If you are envisioning moving to a new program model, try filling out a services map for your current program design and then additional maps for potential future designs.

WHAT DOES YOUR AUDIENCE NEED?	WHAT SERVICES MEET THESE NEEDS?	WHO SHOULD PROVIDE THESE SERVICES AND WHY?
▪ ▪ ▪ ▪ ▪ ▪	▪ ▪ ▪ ▪ ▪	▪ ▪ ▪ ▪ ▪

Appendix E: Potential Audience Segments

To identify your target audience, ask "Who does the program need to reach or engage to accomplish your program and communication objectives?" Keep in mind that you may have more than one target audience based on your objectives.

Consider the following groups to help you identify your audiences:

- Administrators of complementary or similar programs
- Other jurisdictions or local entities
- Community-based organizations
- Faith-based organizations
- Contractors
- Potential program funders
- Students
- Renters
- Landlords
- Utilities
- Volunteers
- Local political leaders and decision makers
- Business owners
- Experts
- Residents
- Neighborhood associations or block groups
- Homeowners associations (e.g., condo boards)
- Universities
- Non-profits
- Green teams or sustainability groups
- Community "gatekeepers" or leaders
- Others

Appendix F: Performance Indicators

Performance indicators measure progress toward program goals and objectives. Both quantitative and qualitative indicators are valuable to track. As you brainstorm, assess, and select indicators, think about the narrative you are hoping to tell. Think about what metrics and data could strengthen the story and make it more compelling for the intended audience.

Brainstorm

Start with a brainstorming session to develop a comprehensive and creative list of potential indicators. Before you brainstorm, review the project goals—all tracking and reporting activities should be directly related to project goals. At the same time, remember that this is a brainstorming session: Be inclusive!

Assess

Now that you have a broad list of indicators, it is time to determine which indicators are valuable and feasible for measuring the success of your project. There are three essential qualities for selecting a good set of indicators. Each indicator must be (1) relevant, (2) measurable, and (3) accessible.

Relevant: Is the indicator useful in determining if the project goals are being met? Is it programmatically important? Is it relevant to the audience(s) you will be sharing your results with? If the indicator does not contribute to understanding the success of meeting project goals, it is not a good allocation of resources.

Measurable: Is it possible to track progress? If the indicator is qualitative, is it possible to rank the evaluation (e.g., high/medium/low or excellent/good/satisfactory/needs improvement) so improvements can be tracked? Does it provide an accurate measure of a task? Can it be defined in clear terms? Will it be consistently measured the same way by different observers?

Accessible: Is the project team able to obtain the necessary data for this indicator at intervals that are appropriate for the project goals? Barriers to accessibility may include data privacy, inadequate resources to collect the data (e.g., staff time, technology), or aggregation of data at too high of a level for it to be useful.

Select

Once a set of viable (relevant, measurable, and accessible) performance indicators are identified, the project team should select the list of indicators that fit the project best. Consider the story you want to be able to tell with your indicators; select a set of indicators that narrate a story of success or precisely guide the implementing body to adjust the program to optimize future success. **Your final list of indicators should be sufficient and succinct.** Too few indicators will provide the project team with limited information. Too many indicators will be burdensome and deter regular tracking and reporting.